Views of the Past

Topographical drawings in the British Library

ANN PAYNE

The British Library

© 1987 The British Library Board

Published by
The British Library
Great Russell Street
London WC1B 3DG

and 27 South Main Street,
Wolfeboro, New Hampshire
03894–2069

British Library Cataloguing in
Publication Data

Views of the past: topographical
drawings in the British Library.
   1. Topographical drawing
   2. Drawing, British
   I. Payne, Ann   II. British Library
   741.941   NC228

ISBN 0 7123 0130 5

Library of Congress Cataloguing in
Publication Data
applied for

Designed by Roger Davies
Typeset in Monophoto Ehrhardt
by August Filmsetting, Haydock,
St Helens
Colour origination by York House
Graphics, Hanwell
Printed in England by
Jolly and Barber Ltd, Rugby

Contents

Introduction

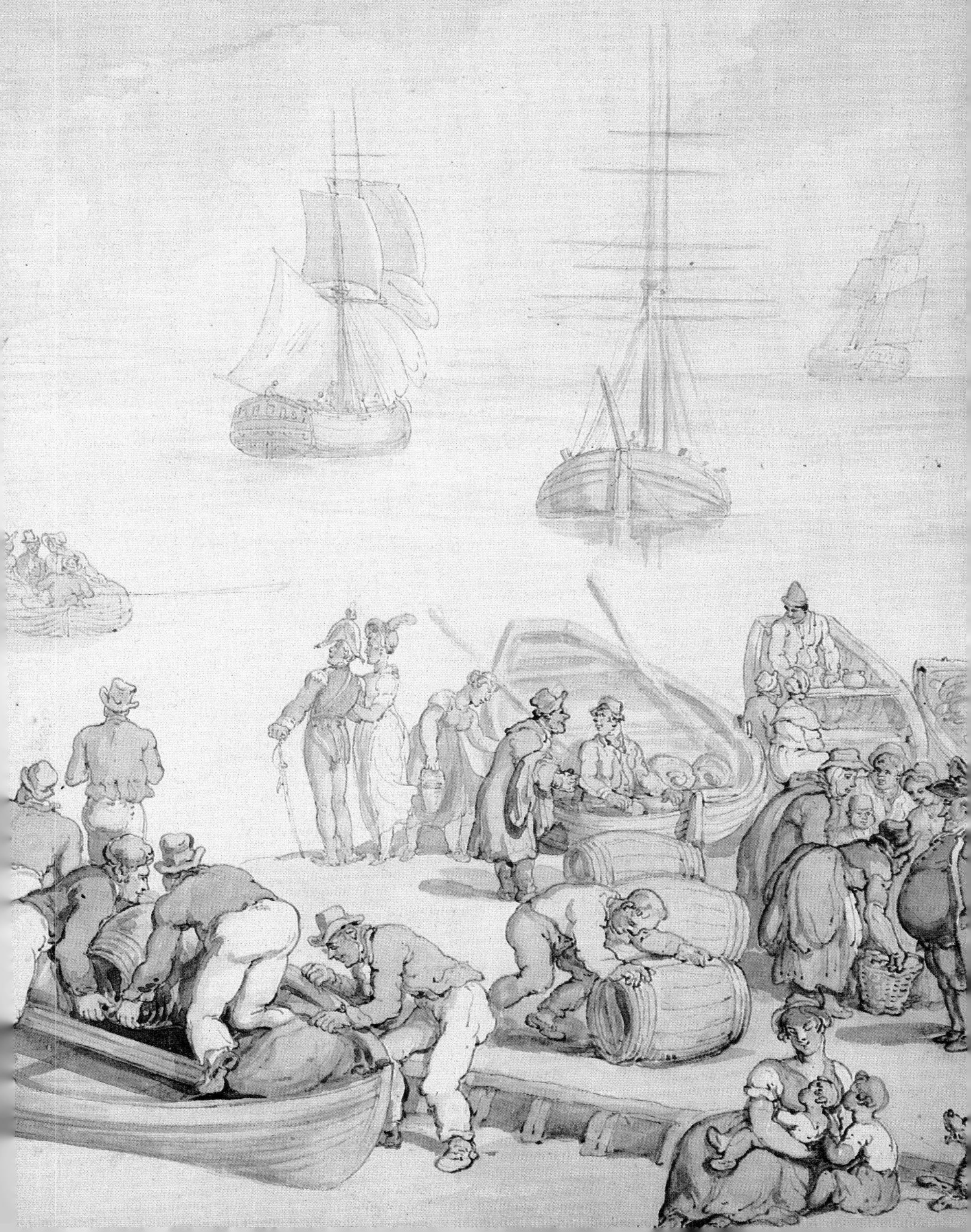

The British Library's department of Manuscripts holds the largest collection of British topographical drawings in the country. The selection illustrated here is an attempt to show the range and variety of this material, supplemented by a few examples chosen from the famous topographical and geographical collection made by George III (the 'King's Topographical collection') which is housed in the Map Library. Choice has been confined to original drawings and watercolours.

Historically the division of drawings of topographical and architectural subjects between the department of Manuscripts and the department of Prints and Drawings in the British Museum has never been altogether clear-cut. Broadly speaking, where the interest is not primarily artistic such drawings have been considered more appropriately retained with the associated written material. In practice the spheres of the two departments have always overlapped. Transfers in both directions have taken place from time to time, but it is still not difficult to find exceptions to any general rule. While it is true that you may look to the Manuscript collections in vain for the masterpieces of British art, this does not preclude the more modest pleasures of discovering unfamiliar works by such artists as Thomas Rowlandson, Edward Dayes, William Mulready and Paul Sandby.

The main strengths of the Library's collections lie in two areas. For the early history of topographical landscape in this country the great series of

1 (*pages 4–5*) The Medway, near Chatham, 1820; Thomas Rowlandson (1756–1827).
Watercolour. 305 × 455 mm. Add. MS 32358, f.129.

2 North East Prospect of Castle Lyon, Linlithgowshire, Scotland, *c.*1746. John Elphinstone, Practitioner Engineer.
Pen, ink and grey wash. 203 × 305 mm. K. Top. L 62.

3 St Albans, Hertfordshire, 1804; William Mulready, R.A. (1786–1863).
Watercolour. 235 × 275 mm
Add. MS 9063, f.128v.

Tudor maps, plans and views formed by Sir Robert Cotton (died 1631) whose manuscripts became one of the foundation collections of the British Museum, is of enormous importance. Secondly, there is a wealth of material for the period 1760 to 1860, reflecting the great age of English watercolour painting and topographical or landscape drawing. This includes a number of large collections of the work of particular artists. Most notable among these for quality of workmanship, geographical spread and sheer quantity are over 3,000 drawings of the social scene of late 18th century England by Swiss-born Samuel Hieronymus Grimm; some 700 drawings by the antiquarian and architect John Carter for his *Ancient Architecture in England*; 45 volumes of monochrome drawings with an emphasis on architecture by Edward Blore; and more than 12,000 drawings by John Buckler, his son John Chessell and his grandson Charles Alban spanning virtually the whole of the 19th century.

Of incidental source material there is a great variety: travel journals and sketchbooks, illustrated reports from voyages of discovery and sightseeing excursions, notes, letters and papers, personal albums and official

4 Rochester, Kent, 1791;
Edward Dayes (1763–1804).

Watercolour over pencil. 142 × 217 mm.
Add. MS 34115, f.7.

5 Sir John Elvil's House on
Englefield Green, near
Egham, Surrey;
Paul Sandby, R.A.
(1730–1809).

Pen, ink and watercolour.
240 × 350 mm. Printed Books, Crach. 1
Tab. 1. b.1.

This drawing is taken from
an extra-illustrated copy of
O. Manning and W. Bray,
*History and Topography of the
County of Surrey* (1819),
enlarged in the mid-19th
century from 3 to 30 folio
volumes with a collection of
prints and drawings formed
by Richard Percival.

commissions. A great deal of material is to be found in historical collections made for proposed county histories or in extra-illustrated copies of such works. The practice of extra-illustrating or 'grangerising' was established in 1769 when the print collector James Granger published a *Biographical History of England* with blank leaves for additional portraits or other pictures. Filling up a 'Granger' became a popular hobby and other suitable texts were soon brought into use. County histories in particular lent themselves to this treatment. The 19th-century vogue for compiling these vast works and the tireless industry of the 'noble band of Grangerites' has left a wonderful quarry. One of the fullest examples in the Manuscripts department is contained in twenty-eight volumes of illustrations collected by a London merchant, J. W. Jones, to enrich histories of Hertfordshire and Kent. In 1881 the British Museum Trustees sanctioned a payment of nearly £1,000, a sizeable sum at that time, to procure these collections: Clutterbuck's *History and Antiquities of Hertford* 'tastefully illustrated' with 1040 additional engravings and 467 drawings, and Hasted's *Kent*, '3710 drawings by eminent artists and nearly 5,000 engravings'.

For the years after about 1860 the material in the Library becomes a good deal scarcer. Topographical drawing is a functional art. Its main purpose is to record accurately; but the invention and development of the camera brought into common use an instrument that was able to record the features of a place with more speed, accuracy and reliability than a pencil and paintbrush.

Thomas Love Peacock found space in *The Misfortunes of Elphin* (1829) to lament the ill-usage of the abbot's kitchen at Glastonbury: 'These ruins were overgrown with the finest ivy in England, till it was, not long since, pulled down by some Vandal, whom the Society of Antiquaries had sent down to make drawings of the walls, which he executed literally, by stripping them bare, that he might draw the wall, and nothing else.' Making exact records of places might on occasion vie with aesthetic discernment. An emphasis on straightforward and sometimes finicky copying, and the non-exclusive participation of an assortment of soldier-artists, local historians, travellers, clergymen, heralds and eager amateurs has consigned topographical drawing to the rank of a humble art. But many men of taste and talent have been lured into topographical work. The drawings in great measure have charm, and nearly always immediacy. Some have more than charm. And even where achievement in terms of artistic merit is slight, the value of the legacy to the student of the past is unquestionable and increasingly widely recognised.

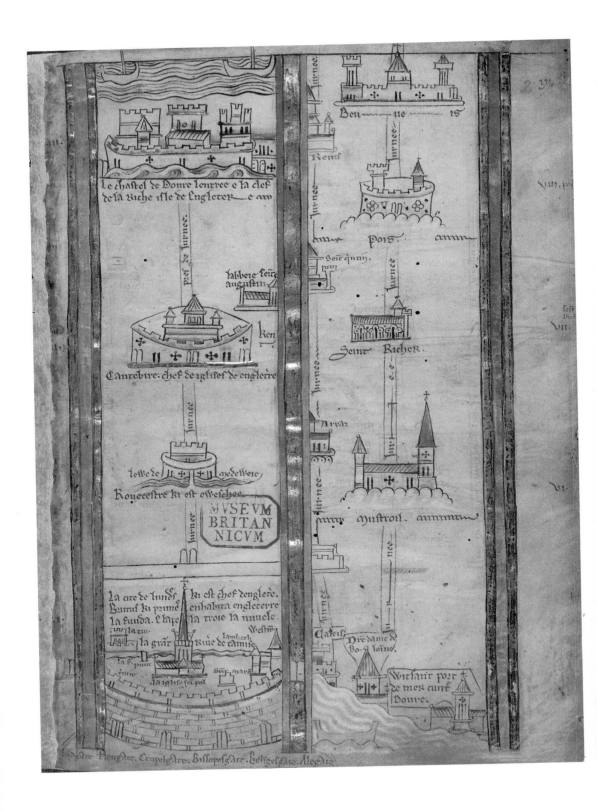

Le chastel de Dovre l entree e la clef
de la riche isle de engleter   e aw

Port de Jurnee

l abbere Seint
augustin

ken

Cantebire. chef de iglises de engletere

Jurnee

lesse de [MUSEUM BRITANNICUM] aue dewere

Rouecestre ki est eveschee

Jurnee

Jurnee

La cite de lundi ki est chef denglere.
Brutus ki primer enhabita engleterre
la funda. e la tu la troie la nunie.
la tu   Westm
la gra Riue de tamise   lambeth
la f punt
entre   Seit mart
la iglise sei pol

Benue   15

Rems

Jurnee

Jurnee

Pois

Seit entin

Jurnee

Saint Richer

Jurnee

Arraz

Jurnee

Muistroil

Jurnee

Caleis   Nre dame de
Bo Loine

Witsant port
de mer cunt
Doure

Early drawings

The scribes and illuminators of pre-Renaissance England usually found it unnecessary in serving their subject matter or their patrons' requirements to introduce into their work pictures of actual places and buildings. Only here and there is it possible to pick out examples which can qualify as genuine topography. One of the most delightful is to be found among the vignettes painted in the margins of the 14th-century Luttrell Psalter (7). It shows a thatched watermill of timber and brick built beside a dammed stream. The carefully executed detail – the boss-headed posts of the dam, the five-spoked wheel, the basketwork eel-traps set in the stream – makes it reasonable to suppose that the picture represents Sir Geoffrey Luttrell's own watermill at Irnham in Lincolnshire.

The 13th-century monk of St Albans, Matthew Paris, embellished his pictorial itineraries with thumbnail sketches of towns passed along the way. Marking the overnight stops on a journey to Rome and Apulia (6) are views of London (seen from the north), Rochester, Canterbury, Dover, and on into France and Italy. Although the form is mostly conventional – standard city walls and towers of an ideogrammatic type used from classical times – one or two of the buildings are almost recognisable. London, for instance, includes old St Paul's with its set of three high windows in the tower and smaller windows above, and with the tall steeple which was eventually to be destroyed by fire in 1561.

Like Matthew Paris's view, the London of the well-known White Tower miniature in the Royal collection (8) is presented from a bird's-eye viewpoint. But whereas the modern observer might be hard-pressed to identify the landmarks of Matthew Paris's London were they not helpfully labelled, there is no mistaking the architectural features of the later drawing. Allowing for some liberties taken with the appearance of the White Tower to accommodate the figure of the prisoner Charles of Orleans, and for a certain distortion in the geography of the River Thames, the artist has concocted a fairly accurate picture of the City. From Traitor's Gate and the Tower we look across the Pool of London with the Custom House on the right to London Bridge and St Paul's beyond. In its competence and its detail it confirms that Italian improvements of the mid-15th century in the technique of aerial views and picture-mapping soon reached Northern Europe. Recent scholarship has shown that this remarkable miniature was made not as previously supposed for Henry VII, but for the Yorkist monarch Edward IV in the early 1480s.

Such bird's-eye views highlight the early and lasting link between topographical and cartographical drawing. Another early influence which was to remain important was the contribution to topography of the heralds. In the middle of an heraldic roll pedigree, made about 1463 to celebrate the Earldom of Salisbury, appears a building which purports to be Bisham Priory in Berkshire (11). The Priory has now gone and no ground plan is known. It is impossible to be certain therefore whether the artist is here attempting to depict a real building and not something invented by his imagination. The inclusion of the carving of the Holy Trinity to which Bisham was dedicated and the tracery of the east window do however

6 (*opposite*) Itinerary from London to Apulia. Matthew Paris, mid-13th century.

Bodycolour on vellum. 368 × 242 mm. Royal MS 14 C. VII, f.2.

te: longe fecit a nobis iniquitates
nostras
Quomodo miseretur pater filiorū
misertus est dominus timentibus
se: quoniam ipse cognouit figmen

7 Irnham Mill in
Lincolnshire.
Marginal illustration in the
Luttrell Psalter, about
1320–40.

Watercolour on vellum. 354 × 244 mm.
(detail).
Add. MS 42130, f.181.

8 A miniature of
the Tower of London with
London Bridge and the City.
From a manuscript of the
poems of Charles d'Orleans,
about 1480–1483.

Bodycolour on vellum. 365 × 270 mm.
(detail)
Royal MS 16 F. ii, f.73.

suggest that he may have intended a real portrait.

By the end of the 15th century evidence of heraldic involvement is clearer. The duties of heralds in connection with arms, genealogies and ceremonial (10) gave them a natural interest in antiquarian matters and in compiling pictorial records. Among the most active in amassing such records was Thomas Wriothesley, Garter King of Arms from 1505 to 1534. A volume of his notes and drawings relating to funerals includes what appear to be the oldest surviving drawings to show not just coats of arms but the tombs, monuments and stained-glass windows which formed their settings. The earliest tomb depicted in the volume is that of Richard Beauchamp, Lord St Amand, who died in 1508 (9). Like the other tombs in the manuscript this one has not survived. Nevertheless there seems little doubt that Wriothesley's drawing of the altar tomb with its elaborate late Gothic canopy was copied direct from the monument itself where it stood in the London church of the Dominicans.

Heralds gave a further impetus to antiquarian and topographical studies by the system of Visitations which operated in England between 1530 and 1686. Kings of Arms or their deputies carried out countrywide heraldic surveys whose main purpose was to regulate the use of arms – 'to correct, deface and take away all manner of arms wrongfully borne or being false

9 Monument to Richard Beauchamp, Lord St Amand (died 1508) in the London church of the Dominicans. Part of the collections made by Sir Thomas Wriothesley, Garter King of Arms 1505–1534.

Pen, brown ink and watercolour on paper. 246 × 335 mm.
Add. MS 45131, f.82.

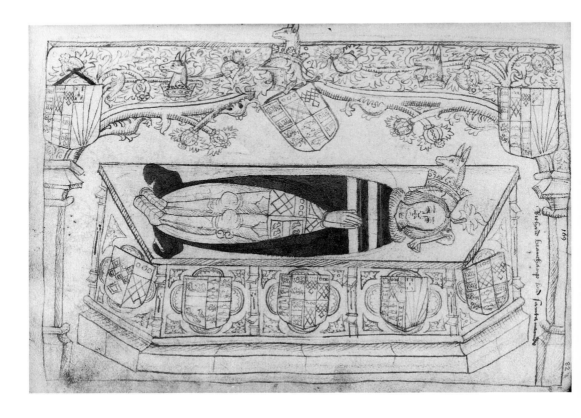

10
Procession leaving the
gateway of the Tower of
London on the eve of Queen
Elizabeth I's coronation, 14
January 1559. The first of a
set of diagrammatic drawings
in the sketchbook of a herald.
Pen and brown ink over black chalk.
395 × 282 mm.
Egerton MS 3320, f.1.

armory'. To this end they added to the pedigrees recorded in their Visit-
ation notebooks, sketches and records of church monuments and inscriptions as evidence of use. The 'Church Notes' of these heralds on their
journeys up and down the country remain a valuable topographical source.

11 (*overleaf top*)
Bisham Priory church,
Berkshire, with its founder
William Montagu, first Earl
of Salisbury and his wife.
From the Salisbury Roll,
about 1463.

Bodycolour on vellum.
Loan MS. 90, p.188 (on loan from the
Duke of Buccleuch and Queensberry,
K. T.).

Military
and
antiquarian
tradition

The British Library's topographical drawings that survive from the 16th century, in particular those of the Cotton collection of Tudor manuscripts, make clear that the principal use found for such draughtsmanship was in the service of national defence. When record keeping was required, instead of written topographical descriptions, surveyors, military planners, builders and, later, landowners turned steadily and increasingly in the course of the 16th century to drawn records. Surveying and engineering practice might in fact make little distinction between topographical prospects drawn from a bird's-eye viewpoint and maps presented, as they most commonly were, in diagrammatic or pictorial form. With improvements in surveying instruments and mapping techniques, and with the growing popularity of the landscape and 'prospect' painting offered by immigrant Dutchmen and Flemings in the mid-17th century, cartography and topography moved further apart. One might say that the vantage point of

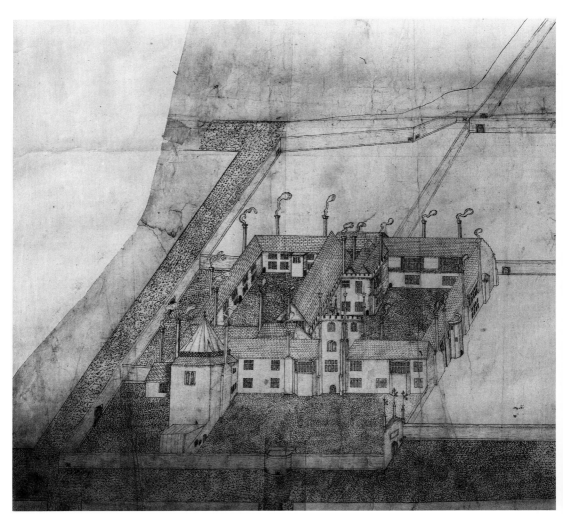

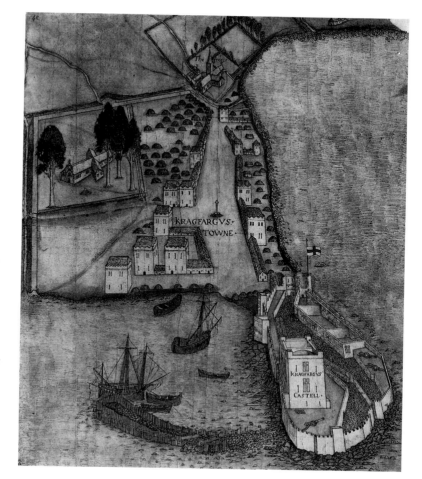

12 (*pages 16–17 below*)
Picture-map of Dover, 1538;
probably drawn by Richard
Lee, surveyor, to illustrate
harbour works.

*Pen, brown ink and watercolour on
vellum. 787 mm × 1.95 m.*
Cotton MS Augustus I.i. 22 & 23.

13 (*left*) Henry VIII's Manor
House at Kingston-upon-
Hull, Yorkshire, 1542–3;
(?) John Rogers, royal
surveyor (died 1558).

Pen and ink on paper. 783 × 770 mm.
Cotton MS. Augustus I.ii.13.

The house was relinquished
by the Crown in 1550 and
demolished in about 1658.

14 (*right*) The castle and
town of Carrickfergus in
Ireland, *c.*1560.

*Pen, ink and watercolour on paper.
660 × 520 mm.* Cotton MS Augustus
I.ii.42.

mappers moved higher and that of most topographers moved lower.

Many of the Cotton maps and bird's-eye views were prepared at royal
command and had a military or naval purpose. One of the most accomplished is a view of Dover harbour (**12**) which it is now thought may have
been the work of the English military engineer, Richard Lee. Drawn in
September 1538, it is part of a group of such 'platts' which date from the
period following the break with Rome when foreign invasion from across
the Channel presented a very real threat. Henry VIII and his military
advisers had need of up-to-date and accurate surveys of the country's
coastal defences. Much emphasis is given in the Dover picture-map to the
lay-out of the newly improved harbour works, and, because the picture
could give only estimated dimensions, it is supplemented with notes and
measurements ('the south est Jette conteneth in lenght viij hundreth and
twenty footes'). The view beyond the distinctive pink-wash sea, showing
Dover town and castle with the cliffs and downs, forms in itself an attract-
ive watercolour landscape.

CANTERBVRY.

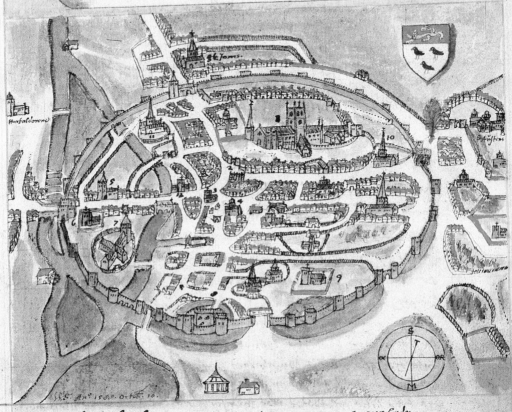

1. christs church.
2. ỹ market place.
3. our Lady.
4. st Andrewes.

5. st peter.
6. westgate church.
7. st mildred.
8. the castell.

9. our Lady.
10. st george.
11. the freeres.
12. Alhalows.

15 (*left*) Canterbury, 1588: one of 15 'portratures' of towns in the 'Particular Description of England . . .' by William Smith (*c*.1550–1618).

Pen, ink and watercolour.
152 × 203 mm.
Sloane MS 2596, f.15.

16 Bird's-eye view of Windsor Castle, 1607; from the 'Description of the Honor of Windesor' by John Norden (1548–1625?).

Pen, brown ink and watercolour on vellum. 410 × 855 mm.
Harley MS 3749, f.3.

Norden, as Surveyor of the Crown woods and forests in Berkshire, executed the manuscript for James I, receiving payment of £200.

A later Tudor example in this tradition of military at-a-glance views of coastal fortifications is the coloured plan or bird's-eye view of Carrick-fergus in Ulster (**14**). Although the view shows interesting topographical details in the town with its market cross and bee-hive shaped huts, it centres on the castle and harbour defences. The main purpose of the drawing was undoubtedly to show their ruinous state – 'so fare in decaye' as the energetic constable of the castle William Peers complained to all and sundry. The complaints (and the drawing?) had their effect in the face of Elizabeth's notoriously miserly bureaucracy: by 1568 the castle had been repaired, the garrison increased and the buildings of 'the freres', the former friary seen to the north-east of the town, converted to a fortified depot.

John Rogers, master mason, military engineer and surveyor of the fortifications at Hull and Berwick, was almost certainly responsible for a striking bird's-eye view of Hull Manor House (**13**), executed in 1542 or 1543. Improvements were proposed to this royal manor, as a result, no doubt, of Henry VIII's visit in the autumn of 1541, and Rogers's drawing may well have been made in this connection.

To match the military surveys Tudor Britain was also being subjected to increasingly close topographical study of a non-military kind. In the wake of John Leland's countrywide explorations of 'things very memorable' in his *Itinerary* of the 1530s and 1540s, educated Tudor gentlemen took to topography in some numbers. The best known work, William Camden's *Britannia* (first published in 1586) covered the whole kingdom, but more often surveys did not reach beyond a single shire. Illustrations were little used in the printed texts until the beginning of the 17th century. Even before that time, however, the advance of topographical scrutiny in county surveys, heraldic visitations and local histories threw up a variety of pictorial material reflecting the interests of those who were undertaking

ECCLESIA MENEVEN

the work: Roman antiquities, pedigrees, coats of arms, churches and their monuments, maps and charts.

From the papers of George Owen of Henllys, Welsh squire and historian of Pembrokeshire, comes the earliest known picture of the cathedral church of St Davids (17), drawn by Owen himself or by one of his servants, its plain and simple lines enlivened by heraldic decoration, a friendly sun and corkscrew clouds.

Much of the topographical work of the 'learned & langwaged' William Smith, a colourful figure who was in his time merchant, traveller, herald and cartographer – even for a while keeper of the Goose Inn in Nuremberg – concentrated on the County Palatine of Chester. But for 'The Particular Description of England . . . 1588', which survives among the manuscripts of the Sloane collections, he cast his sights wider and illustrated the text with 'portratures' or picture maps of the chief cities of England (15).

In similar style and colouring, though done in a surer hand, are the drawings by John Norden of the royal estates in Berkshire made for James I in 1607. His view of Windsor Castle (16) has it laid out in the detailed aerial manner to be expected of a surveyor and proficient map-maker. Norden harboured a grandiose plan for a systematic set of illustrated county surveys to rival Camden's *Britannia* and Saxton's *Atlas*, but he lived to see only Middlesex and Hertfordshire in print.

The engraver Daniel King was associated with the illustrations to a number of topographical works including *The Vale Royall of England* published in 1656. This featured the survey of Cheshire by William Smith and William Webb, and included also 'An Excellent Discourse on the Isle of Man'. Daniel King's drawings for the 'Discourse' are to be found in a sketchbook in the British Library collections (18). They are stiff, uncomplicated and uncoloured, as befitted work whose main function was a guide to the engraver. The addition of key-lettering confirms them as studies for the preparation of book illustration. In the opinion of one of his patrons, the distinguished antiquary and herald Sir William Dugdale, Daniel King was 'a silly fellow . . . an arrant knave'. Certainly he is not noted for the accuracy of his records. Like other British artists of the mid-17th century King was greatly influenced by the Bohemian Wenceslaus Hollar, a map draughtsman of quality, but more importantly a leading exponent of the continental import – ground level prospects and 'landskips'.

On 1 May 1638 William Dugdale joined three other antiquaries, Sir Christopher Hatton, Sir Edward Dering and Sir Thomas Shirley, in forming a society called 'Antiquitas rediviva' (Antiquity restored to life), with an ambitious programme of recording in words and pictures descriptions of medieval tombs, stained glass, manuscripts and armoury. One of the fruits of this association was the 'Book of Monuments' (19, 20). The work was an act of astonishing foresight in face of the threat of approaching Civil War. In the summers of 1640 and 1641 Dugdale took with him an illustrator named William Sedgwick and toured London and the Midlands describing and drawing monuments 'to the end that the memory of them in case of that destruction then imminent might be better preserved for

17 St David's Cathedral, Pembrokeshire, about 1585–6; from the collections of George Owen of Henllys (1552–1613).

Pen and ink. 174 × 267 mm.
Harley MS 6077, f.5.
This is the earliest known illustration of the cathedral.

18 Prospect of Bishopscourt, Isle of Man, about 1656; Daniel King (c.1622–1664).

Pen, brown ink and grey wash.
162 × 273 mm.
Add. MS 27362, f.12.
Engravings from King's drawings of the Isle of Man were published in *The Vale Royall of England*, 1656.

Adhuc Lincoln
Ecclesia cath:

Inter Chorum et alam Aquilonalem

Tumulus Ricardi Fleming
Lincolnensis Episcopi.

19 Tomb of Richard
Fleming, Bishop of
Lincoln (died 1431),
Lincoln Cathedral;
drawn by William
Sedgwick, in Sir
William Dugdale's
Book of Monuments,
1640–41.

*Pen, ink, watercolour and
bodycolour 425 × 280 mm*
Loan MS. 38, f.97 (on loan
from the Trustees of the
Winchilsea Settled
Estates).

20 (*right*) Tomb of Sir
Christopher Hatton
(died 1591), St Paul's
Cathedral; in
Dugdale's book of
Monuments, 1640–41.

*Pen, ink, watercolour and
bodycolour 425 × 280 mm.*
Loan MS. 38, f.186
(on loan from the
Trustees of the
Winchilsea Settled
Estates).

Upon these Spaces are Scituate
Neighbouring Buildings n. front

Some of the
Critchit-fryers Street

Part of Critchit Fryen

21 Sir Christopher Wren's
Navy Office at Crutched
Friars, City of London, 1698.
*Pen, ink and watercolour. 521 × 368 mm
(detail).*
Kings MS 43, f.147.

future and better times'. The results are striking and contain a remarkable
record of now-lost monumental art. Sedgwick may not have been the most
exquisite of draughtsmen and his geometry and perspective were at times
wayward, but his work is bold and colourful, and – a prerequisite in
Dugdale's antiquarian eye – functional. For his later great works, the
Monasticon Anglicanum and his histories of Warwickshire and of St Paul's,
Dugdale employed the talents of Hollar. In almost every case the drawings
illustrating his work are the earliest known of the buildings they record.

In the age of Vauban and the ever-increasing elaboration of city and
harbour fortification, there could be no let-up in the flow of assessments
and reviews by military surveyors. A good example is the work of Thomas
Phillips, master gunner and engineer, whose deft hand and cultured eye
enabled him to turn in some of the most handsome pictorial reports of his
day. Part of Colonel George Legge's 'Present State of Guernsey with a
short Accompt of Iersey' (1680), for which Phillips also produced charts
and soundings ('Truly layd down as they wear survey[d] and Mesured by
T P') were nine watercolour views by Phillips (**23**). These are not aerial
picture maps, but on-site prospects of a number of Channel Island towns

22 Survey Party near Loch Rannoch, Scotland, 1749; Paul Sandby, R.A. (1730–1809).

Pen, ink and watercolour.
172 × 232 mm.
K. Top. L.83–2.

and forts taken from ground level in such an accomplished manner as to confirm that military draughtsmen too had learned from Hollar and artists from the Low Countries.

Slightly later is a 'Survey and Description of the Principal Harbours' of south-east England drawn up for William III in 1698. It is an impressive volume, cataloguing and picturing with some precision naval properties and recently improved dockyards in the Thames estuary and along the south coast, with large prospect views, by an unidentified hand, of, among others, Chatham, Portsmouth and Plymouth. It closes with a neat coloured drawing of the Navy Office itself, off 'Crutchit Fryers Street' in the City of London (**21**). This was erected in 1683–4 to Sir Christopher Wren's design, to replace the building where Samuel Pepys had worked and lodged and which was burned down only a half dozen years after surviving the Great Fire.

It was accepted by military and naval authorities that a reconnaissance drawing of a place made on the spot provided more reliable intelligence than any written accounts. Soldiers and sailors who were able to draw accurately were therefore of great value not only for the preparation of essential maps, charts and elevations but for a lengthening list of functions ranging from the depiction of battle scenes to the illustration of instruction

The Prospect of Mont Orgueill. taken from the hill. Anno 1680 ff Tho Phillips

manuals. Samuel Pepys was among those who, at a time of rapid naval expansion, put their weight behind schemes to improve the quality of naval draughtsmanship. Pepys was prominent in the establishment in 1673 of the Royal Mathematical School at Christ's Hospital, whose purpose was the training of boys for the navy. Here the syllabus came to include not only the science of navigation and mapping but also 'the practice of Drawing for laying down the appearance of Lands, Towns and other objects of notice'. Pepys and Sir Christopher Wren also supported the proposal of the Christ's Hospital governors in 1692 to introduce drawing-masters into the school and from 1705 the 'Mathemats' were given the benefit of their own professional Drawing Master. The Royal Military Academy at Woolwich acquired its first such master in 1743, and its most distinguished one, Paul Sandby, in 1768. Sandby held the post for almost thirty years and consequently was in a position to play as important a part in the development of military topography into the 19th century as he did in that of English watercolour painting. An unfinished drawing of the Royal Foundry (24) is attributed to Sandby, done during his time as an army drawing master at its Woolwich neighbour, the Academy.

Paul Sandby and his older brother Thomas were from a cartographic background. Both had been attached as Board of Ordnance mappers to the forces of the Duke of Cumberland in Scotland after the 1745 Jacobite Rebellion. Paul became the official draughtsman of a team led by William

23 Castle of Mont Orgueil, Jersey, 1680; Thomas Phillips (?1635–1693), a royal engineer.

Pen, pencil and watercolour.
710 × 412 mm.
Kings MS 48, ff.74v–75.

24 The Foundry, Woolwich, n.d.; attributed to Paul Sandby, R.A. (1730–1809).

Pencil, Indian ink, wash and watercolour. 311 × 470 mm. Add. MS. 32375, ff.105v–106.

Roy making a detailed survey of Scotland, which was thought desirable as part of a policy of strengthening central control. His hand is seen in many of the original maps of the survey now in the British Library, and he also produced a splendid illustration of the surveying party at work in the Grampian Highlands (**22**).

Though a recognisable enough figure in earlier times as a man who would go 'forty miles to see a Saints well or a ruin'd Abbey', the anti-quarian topographer enjoyed his real heyday in the eighteenth century. From 1717 the newly re-established Society of Antiquaries sponsored the publication of fine engravings and its members individually acted as en-thusiastic patrons and collectors. The topographers *par excellence* who catered to this growing market of the learned and curious were the bro-thers Samuel and Nathaniel Buck.

The British Library possesses Samuel's earliest surviving work, a sketchbook of his home county, Yorkshire, made on a series of tours between 1719 and 1723 (**25**). These rough pen-and-ink sketches of country houses and town prospects were commissioned by John Warburton, excise man, antiquary and herald, whose grand scheme was to produce a county history of Yorkshire. Warburton's project proved over-ambitious and came to nothing; but Samuel Buck was able to use his sketches in later enterprises and his notebook remains an invaluable architectural record.

Between 1720 and 1753 the Buck brothers undertook a vast nationwide project to illustrate 'the venerable remains of above 400 Castles, Monas-

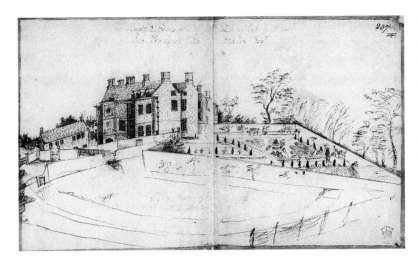

25 Lartington Hall, Yorkshire, *c.*1723; from an early sketchbook of Samuel Buck (1696–1779).

Pen and brown ink. 200 × 310 mm. Landsowne MS 914, f.241.

Formal gardens such as these were soon to be swept away by the landscape garden movement of the mid-18th century.

teries, Palaces, etc, in England and Wales.' They also produced no less than 83 town prospects. In this they had the assistance of a number of other artists who were rather more skilful than they themselves. Summers were spent in travelling and drawing, winter in engraving and spring in publishing. The works issued in this instalment fashion were later published in collected form as *Buck's Antiquities* (1774). Even today the engravings are among the best known of all topographical studies, familiar from modern reprints on sale at sites of historic monuments throughout Britain. As might be expected, the volumes of extra-illustrated county histories in the Library are plentifully supplied with Buck engravings. Less well known are the occasional preparatory drawings in ink and wash: the collection of illustrations to Hasted's *History of Kent*, for example, includes drawings of Tunbridge Priory, and of the castles of Walmer, Saltwood and Leeds (**26**).

Buck's *Antiquities* was one of the early symptoms of the 'Gothic revival' which was to spread through the country from the middle of the 18th

26 Leeds Castle, Kent, 1735; sketch for engraving by Samuel and Nathaniel (fl.1727–1753) Buck .
Pen and Indian ink wash.
162 × 348 mm.
Add. MS 32366, f.163.

27 Thames Ditton Church, Surrey, August 1766; Francis Grose, F.S.A. (1731–1791).

Pen, ink and watercolour.
267 × 375 mm.
Printed Books, Crach. I. Tab. 1. b.1 (vol. VII).

century. Its effect was to convert public taste from the orderliness of classical balance and symmetry to the wild, the ruinous and the romantic, and it made its mark in all branches of art from painting to landscape gardening. Among the fashions it inevitably banished was the kind of estate portrait familiar from the engraved work of such men as the Dutchmen Knyff and Kip: comprehensive and formally-drawn picture maps, which gave aerial prospects of the substance of a gentleman's acres – his tidy house, his straight, level walks and his symmetrical gardens. A watercolour of Campden House in Gloucestershire by the English amateur William Hughes, done about 1750, is a naive but pretty example of the style (**28**). This is a reconstruction by Hughes in 18th century dress, for, while most of the other buildings still stand, the big house itself was in fact burned down by Royalists in 1645. New preferences rebelled not only against the formality of subject of such portraits but also against their 'cartographic' presentation: better now for the old manor house to be espied across picturesque parkland from between unregimented trees.

The antiquarian topographers played a leading role in promoting 'Gothic' appeal. Conspicuous in their ranks were Captain Francis Grose of the Hampshire and Surrey Militias, for a time Richmond Herald, a rotund and jovial celebrity of the Society of Antiquaries whose popular *Antiquities*

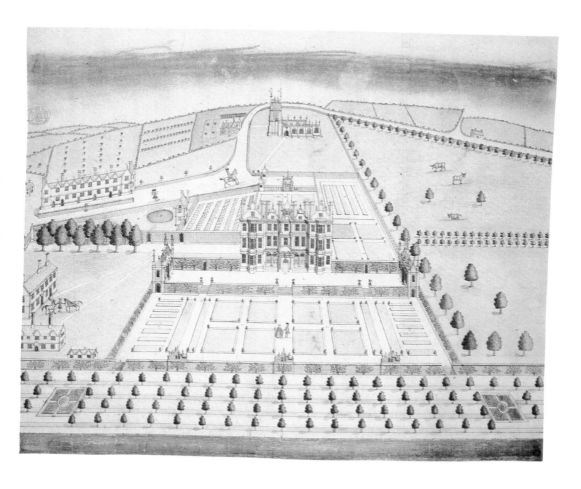

28 Campden House, Gloucestershire, *c.*1750; William Hughes.

Pen, ink and watercolour.
443 × 474 mm.
K.Top. xiii.75.3.

The main building is a reconstruction by Hughes of Sir Baptist Hicks's Campden House which was a casualty of the Civil War: most of the surrounding buildings still stand.

of England and Wales published between 1773 and 1787 was lavishly ill-ustrated with the work of several artists, including his own drawings (**27**) and those of his valet Thomas Cocking; and John Carter, surveyor and architectural historian, whose conscientious study and hectoring cham-pionship of medieval buildings (**29**), helped to bring some historical sense into the muddled popular notion of 'gothic'.

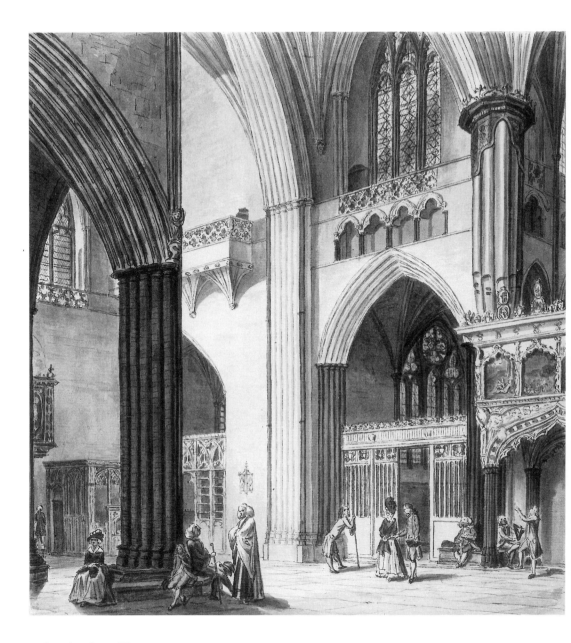

29 Interior view of Exeter
Cathedral, 1770; John Carter,
F.S.A. (1748–1817).

Pen, ink and watercolour.
272 × 240 mm.
Add. MS 29925, f.8.

30, 31 Letheringham Church, Suffolk, 1712 and about 1790.

Pen, brown ink and watercolour. 313 × 194 mm (page size). Pen, ink and grey wash. 250 × 360 mm.
Add. MSS 33247, f.261; 8987, f.87.

The first of these drawings is taken from a page of the 'History of Framlingham' by Robert Hawes, Steward of the Framlingham and Saxted Estates (1665–1731). The late 18th century drawing, by William and Isaac Johnson (1754–1835), shows how far the church had by then fallen into decay; the fine monuments of the Wingfield family seen here in a derelict state were subsequently to fall prey to the attentions of collectors and sale by the contractor engaged to rebuild the church.

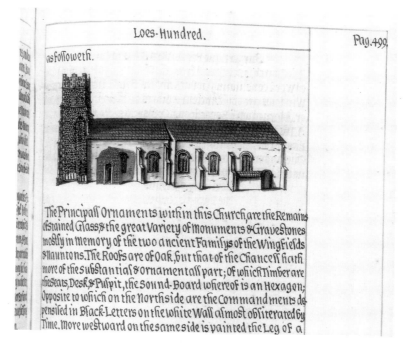

'Itinerant
View
Takers'

33 The Resident's House at
Southwell, Nottinghamshire,
1787; Samuel Hieronymous
Grimm (1733–94)
Pen, Indian ink and wash.
270 × 376 mm.
Add. MS 15544, f.173.
Grimm's patron Dr Richard
Kaye was for a period
Prebend of Southwell and
Grimm drew the Cathedral
and the houses in the close
from every aspect.

32 (*pages 36–37*) Fountains
Abbey, Yorkshire; Samuel
Hieronymous Grimm
(1733–94).
Pen, ink and watercolour.
378 × 545 mm.
Add. MS 15548, f.49.

The lawyer Sir William Burrell, planning a history of Sussex which he never completed, commissioned over the period 1780 to 1791 a series of illustrative drawings from James Lambert, a local watercolourist (see **front cover** and **35**), and from Samuel Hieronymus Grimm, an immigrant from Switzerland. The ecclesiastic Sir Richard Kaye made use of Grimm to produce countless drawings of English scenes and antiquities, largely for the fun of collecting the work as records of their excursions together in over half the counties of England. Both collections of drawings, totalling twenty-one volumes, were bequeathed to the British Museum and are now together in the department of Manuscripts. The vast collection of Grimm drawings is of enormous historical interest. His brief from Kaye was topographical in its broadest sense – 'everything curious' that they came across in their travels: houses and churches, gateways and wells, castle and ruins, people at work and at play. Most of the drawings are in ink and wash (**33, 34**), but there are a few large coloured views, as for example that of Fountains Abbey (**32**), a favourite subject among disciples of the 'picturesque'.

Commissions for Burrell, Kaye, the naturalist Gilbert White and others involved Grimm in almost constant travel across all parts of the country to an extent that would probably have been unendurable in a previous age. A gradual improvement in roadmaking and in road transport greatly reduced much of the dreadful discomfort that had long been associated with

34 A service in Bath Abbey, 1788; Samuel Hieronymous Grimm (1733–94).

Pen, Indian ink and wash.
380 × 540 mm.
Add. MS 15546, f.101.

Arundel Castle 1783

36 (*above*) Wentworth House, Yorkshire, 1774; (?) Theodosius Forrest, R.A. (1728–1784) after Thomas Sandby, R.A. (1723–1798).

Pen, ink and watercolour.
265 × 370 mm.
Add. MS 42232, f.40.

35 (*left*) Arundel Castle, Sussex, 1783; James Lambert (1725–1788).

Pen, brush and watercolour.
305 × 242 mm.
Add. MS 5677, f.76.

journeys. In the 18th century people took to travelling as a tolerably comfortable leisure activity. Where once they had travelled only when necessary, now they began to 'tour'. Growing numbers of the nobility and gentry also began to buy drawing tuition. Drawing masters became available in almost every part of the country. Whole families were given instruction in the principles and techniques of making pictures in pencil or watercolour, and went out to practise the craft. An accompanying expansion took place in the trade in topographical prints, sold as portfolio collections or as the illustrations to travel books and guides. Travellers liked to be shown what they were to look at and liked to have souvenirs of what they had seen. To satisfy a hungry market, not only were professional painters dispatched about the countryside in their droves, but amateur work too was taken up by publishers to be turned into saleable prints. At times, as the 18th century turned to the 19th, it must have seemed as if every artist in England, the trained and the half-trained, was out and about, sketchbook in hand, criss-crossing the land in search of the 'picturesque' view for personal albums or for publishers' profits.

An agreeable example of an artist's illustrated tour journal has been recently identified as a fair copy – the original is in private hands – of Thomas Sandby's excursion round Yorkshire and Derbyshire in August 1774 with a party that included his artist friend Theodosius Forrest. The manuscript is included in a volume of three journals having the collective

37 Silhouette of Thomas
Sandby, R.A. (1723–1798)

107 × 87 mm.
Add. MS 36694, f.11.

title 'Forest's Tours', and seems likely to have been a copy made for
Forrest himself. The watercolour pictures illustrating the tour are so close
to the originals that either they must be by Sandby or very precise copies
by Forrest, whose own style was rather similar to Sandby's. Sandby's
account is in the form of letters to his wife: 'As I wish to give you a pretty
good idea of everything we meet with on Our tour worth noticing I shall
make Sketches of such Buildings as please me most for your inspection
that you may the less regret not being one of the party'. As well as the
'pleasing sights' he dwells also on the discomforts of stage coach travel –
the bad food, uncivil landlords, even on one occasion the linen unchanged
from previous travellers. A high point was the visit to Wentworth House,
near Sheffield (**36**). It was usual for the nobility to open its parks and
mansions to presentable visitors, but Lady Rockingham's hospitality was
exceptional: 'chit chat' audiences with her ladyship, good dinners and beds
for the night in her palatial house; 'Tho' it seem incredible yet I will
venture to say, I was led at least Two thousand feet to the room appointed
for my repose'.

Thomas Rowlandson is thought of as a caricaturist rather than a topographer – too busy with the liveliness of a scene to focus on topographical
and architectural niceties. However his prolific output included a great
many prospects and landscapes. The British Library has handsome
examples of his work in views of St. Albans (**39**), Taplow in Buckinghamshire (**40**) and the Medway near Chatham (**1**), combining to show that for
Rowlandson topography was invariably populated. A free-living, con-
vivial, energetic man, he was not only a regular tourist himself in Britain
and on the continent, but had great fun in mocking the 'picturesque tour'
in his Dr Syntax series of drawings.

One of Rowlandson's cronies and travelling companions was Henry
Wigstead, an amateur whose artistic aspirations so outstripped a modest
talent that the joint hoax the friends played on the Royal Academy in
passing off some good watercolours by Rowlandson as the work of
Wigstead could hardly stay long unexposed. In a volume of drawings in
the British Library by Wigstead, Rowlandson and others is a set prepared
to illustrate a tour of Kent made about 1797. The monochrome pictures
stand ready for the printmaker but the tour was never in fact published.
Rowlandson had helped to illustrate Wigstead's *Remarks on a Tour to
North and South Wales* after the two had explored the principality
together, but the helpful expert in the present case, ready to supplement
and improve Wigstead's own efforts, was almost certainly Rowlandson's
brother-in-law Samuel Howitt.

In 1799 the artist and traveller Robert Ker Porter, then a young man of
twenty-two, helped to found a 'select society of Young Painters' with the
avowed intention of practising 'Historic Landscape'. Meetings of the
society, which numbered Thomas Girtin among its members, took place
in Ker Porter's rooms in Great Newport Street, near Leicester Square,
once the studio of Sir Joshua Reynolds. The nucleus of the group may
have been formed earlier. Two years previously the secretary, Louis Fran-

cia, and the treasurer, John Denham, appear to have been Ker Porter's companions on a tour of Hampshire and the Isle of Wight. They figure as 'Itinerant View Takers' (43) in the drawing which acts as frontispiece to Ker Porter's sketchbook of the tour. Although Ker Porter's contemporary artistic reputation rested primarily on his later work as a historical painter (at one time by appointment to the Tsar of Russia), and he is now better known as an international traveller, his early sketchbook shows a talent for splendidly atmospheric watercolours of English scenes (44, 45).

The naturalist Joseph Banks was wedded to the idea that a picture is worth pages of words in recording information and descriptions of scenes and landmarks. After his fruitful participation in Captain Cook's first voyage, Banks fell out with the Navy Board and withdrew from the second, preferring instead a foray into closer waters – to the Hebrides, Orkneys and Iceland. He took with him the trio of draughtsmen already appointed for the Cook voyage: John Cleveley, Jnr., James Miller and John Miller. Four volumes of the sketches and watercolours made during this trip of 1772 were bequeathed by Banks to the British Museum, along with a collection of drawings done on the earlier Cook expedition. Banks's faith in pictorial records was rewarded with a set of fascinating drawings, like those which show a hunting party on the 'bending pillars'

38 Verse letter of Thomas Sandby to Theodosius Forrest with a sketch of (?) Clay Hall in Old Windsor where Sandby and his wife lodged; c.1755.

Pen, ink and watercolour.
302 × 187 mm.
Add. MS 36994, f.2v.

39 (*above left*) St Albans,
Hertfordshire, *c*.1795–1800;
attributed to Thomas
Rowlandson (1756–1827).

Pen, ink and watercolour.
278 × 431 mm.
Add. MS 9063, f.138v.

40 (*below left*)
Taplow, Buckinghamshire,
c.1795–1800; Thomas
Rowlandson.

Pen, ink and watercolour over pencil.
127 × 206 mm.
Add. MS 18674, f.16.

41 The Abbot's Kitchen at
Glastonbury, Somerset, 1797;
Rev. John Skinner
(1772–1839), rector of
Camerton, near Bath.

Pencil and watercolour. 115 × 168 mm.
Add. MS 33635, f.57.

John Skinner is one of the
best known of English
country parsons whose leisure
was devoted to topographical
and antiquarian pursuits.
Over one hundred volumes
of his journals in the
department of Manuscripts
show him to have been an
indefatigable traveller,
especially in the southern
counties, illustrating his tours
with naive watercolour
sketches of places visited and
antiquities discovered.

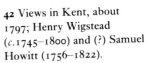

42 Views in Kent, about 1797; Henry Wigstead (*c*.1745–1800) and (?) Samuel Howitt (1756–1822).

Ink and watercolour. 105 × 174 mm. (each view)
Add. MS 18674, f.40.

The drawing of Reculver (top) shows the twin towers of the Saxon church before they were destroyed.

Reculvers

Rottendean

43 'Itinerant View Takers', frontispiece to a sketchbook of a tour to Hampshire and the Isle of Wight, 1797, by Robert Ker Porter (1777–1842).

Pen, ink and watercolour.
325 × 257 mm.
Add. MS 18283, f.1.

Three of the artists are identified by faint pencil inscriptions as J. C. Denham (*fl.*1796–1858), L. T. Francia (1772–1839), and 'Bob', presumably Ker Porter himself; the figure with the brush or cane remains unidentified.

near Fingal's Cave on Staffa (**46**) and members of his team examining curious tombstones at Killaru on Islay (**47**).

Two of the Library's largest individual collections, by Edward Blore and by the Bucklers, are the work of professional architects of the 19th century who must have devoted as much of their time to the drawing of buildings that were, as to the designing of buildings that might be. The Buckler family's huge collection of drawings by its three generations John (1770–1851), John Chessell (1793–1894) and Charles Alban (1824–1904) is in the department of Manuscripts. Extensive enough to rank as perhaps the topographical collections most commonly consulted for reference and research, the emphasis is decidedly architectural. In contrast to Rowlandson or Samuel Grimm, the Bucklers introduced the animate into their

a view of Southampton, from Mr Plews House. Spring Place 30th June 1797 R·K·P

44 Southampton, 30 June
1791; sketchbook of Robert
Ker Porter (1777–1842).

Pencil and watercolour. 325 × 257 mm.
Add. MS 18283, f.11.

45 Carisbrooke Castle, Isle of
Wight, July 1797; sketchbook
of Robert Ker Porter
(1777–1842).

Pencil and watercolour. 325 × 257 mm.
Add. MS 18283, f.26.

46 The 'bending pillars' near Fingal's Cave, the Hebrides, 1772; John Cleveley the younger. (1747–1786).

Pen, brush and Indian ink wash.
322 × 490 mm.
Add. MS 15510, f.34.

pictures only with reluctance. The work is mostly in pencil. Of the relatively small number of watercolours, those by J. C. Buckler are especially pleasing (**48**). His father's tinted interior of Stourhead in Wiltshire (**51**) shows the home of an important patron, Sir Richard Colt Hoare. Charles had less natural talent than his predecessors, but his view of Lanthony Priory in Monmouthshire presents him at his best (**52**).

Edward Blore was also an architect with strong antiquarian interests. He produced a host of illustrations for county and local histories and these feature among the 4,500 drawings which the Library holds. The drawings, typically in pencil or more impressively in sepia monochrome, are done with an eye for detail and a delightful talent for depicting the effects of light and shade on buildings. They range countrywide and include abbeys, churches and public buildings, castles and mansions, landscapes, townscapes and feats of civil engineering like the bridge over the Wear at Sunderland (**50, 53, 55**).

While many artists were dispersing their energies over every corner of the country, there were others who became specialists in portraying particular areas: James Lambert, for instance, in Sussex; James Ross, a Worcester engraver with a nice line in prettily-posed scenes with a touch of caricature, in the neighbourhood of the Severn Valley (**56**); the Cromes in Norfolk (**57**); and William Green in the Lake District. In keeping with a topographical tradition, William Green moved from surveying and mapmaking into landscape drawing, proclaiming as the purpose of his

47 Members of Joseph
Banks's party at the ruins of
Killaru, Isle of Islay, 1772;
John Cleveley the younger
(1747–1786).

Pen, ink and watercolour.
280 × 404 mm. (detail)
Add. MS 15509, f.11.

pictures 'to save from the wreck of time and the busy hand of man' the
'mountain architecture' of the Lake District, his adopted home after 1800.
With the forthrightness of an indefatigably industrious Manchester surveyor having little time for the artistic licence of the picturesque school, he
drew and painted the lakes just as he saw them. His rather sombre, still,
realistic watercolours and aquatints promoted tourism throughout the
Lake District and exploited the growing demands for souvenirs that it
brought with it (**54**).

The Bavarian-born artist George Scharf settled in London in 1816 and
proceeded to illustrate the social topography of the capital with charm,
flair and affection. He was attracted by its busy street scenes and markets,
the changing face of the buildings and the whole workaday life of the city.
Many of these London drawings are now to be found in the British
Museum's Prints and Drawings department, but a smaller group of his
work – some of them from the more personal albums of Scharf and his
family – are among the delights of the department of Manuscripts (**60, 61,
63**).

In the Yarmouth household of Charles Dawson Turner local topographical drawing was made something of a family business. A banker by
profession, Turner had many cultivated interests – as botanist, classicist,

49 (*above*) Hatfield House, Hertfordshire, 1812; John Buckler (1770–1851).

Pencil and watercolour. 208 × 406 mm. Add. MS 32349, f.164.

48 (*right*) Geddington Cross, Northamptonshire, undated; John Chessell Buckler (1793–1894).

Watercolour. 522 × 367 mm. Add. MS 15966. A.

51 The Dining Room at
Stourhead, Wiltshire, 24 May
1824; John Buckler
(1770–1851).

Pencil and watercolour. 265 × 365 mm.
Add. MS 36392, f.155.

50 Durham Cathedral,
about 1823, Edward Blore
(1789–1879).

*Pencil, pen and brown wash with white
bodycolour. 320 × 215 mm.*
Add. MS 42016, no.1.

collector of manuscripts, patron of the arts, and antiquary. One of his projects, the extra-illustration of Blomefield's *History of Norfolk* with, among other pieces, some 4,000 drawings and 3,000 etchings, is now in the British Library. Many of the illustrations are by Turner's wife and five daughters, but they also collected drawings by other Norfolk artists. John Sell Cotman was the family drawing master from 1811 to 1823 and Turner acknowledged his crucial influence in the work ('Whatever merit may be found in this collection of drawings, is mainly attributable to him') – although, rather surprisingly, actual examples of Cotman's original work in it are few (**58**). It was Cotman's own fondness for drawing architectural antiquities that largely inspired the scheme as a family project, and the results of his teaching are apparent throughout. Many of the drawings are slavish imitations, often direct copies, of Cotman's work. The youngest Turner girls managed to develop, after Cotman had left them, a rather more individual style and the drawings of Mary Ann and Harriet (**59, 62**) are perhaps the pick, done with controlled, trim penwork and sometimes pretty colouring. Over a period of thirty years the family worked to complete its visual record of Norfolk and its heritage. Visitors were always impressed by their industry. 'Mrs Turner has been etching with her daughters in the parlour every morning this week at half past six!' reported

52 (*above*) Lanthony Priory,
Monmouthshire, undated;
Charles Alban Buckler
(1824–1904).
*Watercolour and bodycolour with gum
arabic. 276 × 408 mm.*
Add. MS 34115, f.30.

53 (*above right*)
Bamburgh mill and castle,
Northumberland, undated;
Edward Blore (1789–1879).
*Pencil, pen and brown wash.
225 × 313 mm.*
Add. MS 42016, no.41.

54 (*below right*)
Overbeck Bridge on Wast
Water, in the parish of St
Bees, Cumberland, undated;
William Green (1760–1823).
Watercolour. 205 × 305 mm.
K.Top.x.68.

55 (*pages 58–59*)
Sunderland Bridge,
Durham, before
1816; Edward Blore
(1789–1879).
Pencil, pen and brown wash.
158 × 230 mm.
Add. MS 42016, no. 13.

56 (*pages 60–61*)
Northwick Park,
Worcestershire, 1783;
James Ross
(1745–1821).
Watercolour. 381 × 262 mm.
K. Top. XL.III.77.
A note on the back of
the drawing says that
the viewpoint was
chosen by Sir John
Rushout,
Northwick's owner,
but was objected to
by Lady Rushout
because the end of
the water is seen.

57 (*above left*) St Etheldreda Church, Norwich, 1813; John Berney Crome (1794–1842).
Pencil, pen and grey wash.
140 × 200 mm.
Add. MS 23039, f.44.

58 (*below left*) Haynford Church, Norfolk, 11 May 1815; John Sell Cotman (1782–1842).
Pencil, pen and ink. 229 × 291 mm.
Add. MS 23031, f.94.
This church is now in ruins.

59 (*above*) Houses on Laughing Image Corner, Yarmouth, Norfolk, *c.*1830; Mary Ann Turner, daughter of the antiquary and botanist Dawson Turner.
Pen, ink and watercolour.
310 × 229 mm.
Add. MS 23050, f.111.

geologist Charles Lyell admiringly in 1817. Sir Frederic Madden, the British Museum's Keeper of Manuscripts, however, upon first acquaintance in 1832, found the régime less congenial: 'Mr T has 3 unmarried daughters at home, all of whom are extremely clever, and brought up in a systematic plan adopted by Mr Turner for getting up at 7 or earlier, breakfasting at 8, drawing or studying all day, teaching at schools, etc., dining at 5, and going to bed at 10. I confess, that the hours are very irksome to me. . . . I am bored to death'. Within four days, however, the doleful Madden found himself won over, and was half in love with Hannah Turner.

60 The Castle Inn in the Old
Kent Road, 1827; George
Scharf (1788–1860).

Pen, Indian ink and watercolour.
100 × 141 mm.
Add. MS 36489 A, f.20.

61 On the River Thames,
Bermondsey, May 1827;
George Scharf (1788–1860).

Pen, Indian ink and watercolour.
228 × 142 mm.
Add. MS 36489 A, f.108.

G. Scharf del. May 1827.   on the River Thames, Bermondsey

63 'View taken from N. 19
Bloomsbury Sqʳ., London,'
undated; George Scharf
(1788–1860).

Pencil. 142 × 227 mm.
Add. MS 36489 A, f.88.

62 Interior of Burgh Church,
Norfolk *c*.1840; Harriet Gunn
née Turner (1806–1869).

Pencil, pen and wash. 310 × 229 mm.
Add. MS 23026, f.146.

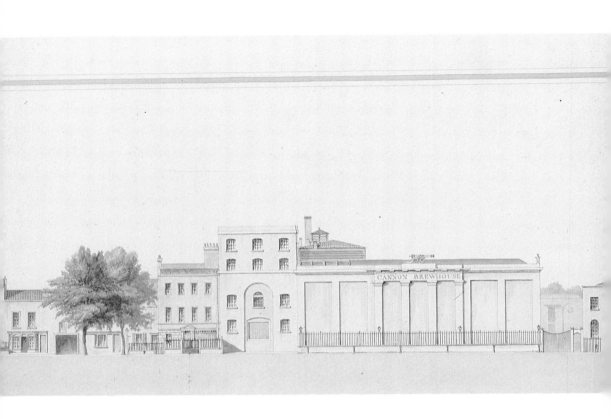

64 Knightsbridge, London, 1811; Joseph Salway, surveyor to the Kensington Turnpike Trust.

Pen, ink and watercolour. 178 mm × 1.14 m (detail). Add. MS 31325, no.12.

This drawing comes from three volumes of plans and views of the main roads for which the Trust was responsible from Hyde Park corner to Counter's, Fulham and Battersea Bridges. Many of the buildings seen here, including the Cannon Brewhouse, were pulled down in 1841.

Travellers abroad

65 Boston from Beacon Hill, 1775; Lieut. Richard Williams of the Royal Welsh Fusiliers. Part of a series of panoramic views around Boston.

Pen, ink and watercolour.
170 × 480 mm.
K.Top. CXX. 38f.

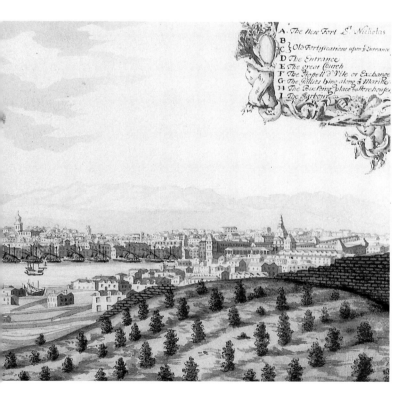

66 View of Marseille, 1685; from 'A Voyage into the Mediterranean seas' by Edward Dummer who became Surveyor of the Navy in 1692.

Pen, brown ink, grey wash and watercolour. 700 × 247 mm (page size)
Kings MS 40, ff.59v–60.

In making drawings of the places they have visited Britons abroad have not been different people from Britons at home. Their subjects might be more unfamiliar and exotic: all the more reason, then, to bring home convincing pictorial records of their discoveries. Wherever military adventures, the expansion of trade and empire, overseas exploration or mere tours of pleasure have taken them, a sketchbook has regularly been a necessary item of baggage.

When HMS *Woolwich* sailed for the Mediterranean in 1682 she had on board as 'Midshipman Extraordinary' Edward Dummer, later to become Surveyor to the Navy. His report on his two-year expedition ('by way of a journal') not only includes descriptions and ingenious paper models of foreign vessels but is also illustrated with over a hundred large-scale composite plans and views of foreign cities, towns, harbour forts and arsenals (66) – military and naval intelligence about European rivals of enormous value to the Admiralty planners who had sent him.

Few military draughtsmen could achieve the stature of Thomas Sandby, whose drawings of encampments in the Netherlands made while on the Duke of Cumberland's staff in 1748 are the work of a skilled artist (67). But the drawings of less technically competent practitioners can at least have the fascination of immediacy and personal involvement. The King's Topographical collection in the Map Library is rich in the work of soldier-artists. A series of panoramic views of Boston and the surrounding country records dramatic events at the outbreak of the American War of Independence. In his simple wash drawings Lieutenant Williams of the

67 (*below*) Panoramic view across country to Zeeland, 22 June 1748; Thomas Sandby (1723–1798).
Pen, ink and watercolour washes. 275 mm × 1.7 metres (detail).
Add. MS 15968 C.

68 (*right*) Vaitepiha Bay, Tahiti, 1777; John Webber (*c.*1750–1793).
Pen, brown wash and watercolour. 448 × 634 mm.
Add. MS 15513, f.13.

69 (*below right*) The *Resolution* in Resolution Cove, Bligh Island, Nootka Sound, 1778; John Webber.
Watercolour. 480 × 660 mm.
Add. MS 15514, f.10.

Royal Welsh Fusiliers has given us a vivid eye-witness impression of the city of Boston and the positions of the besieging Rebels as seen from the vantage point of Beacon Hill (**65**).

Those 'skilled in Drawing' are known to have accompanied colonising expeditions and voyages of discovery from at least the middle years of the 16th century. Joseph Banks's team of artists and scientists on Captain Cook's *Endeavour* in 1769 were not establishing a new practice. Nevertheless the scale of the venture was new. Banks's appreciation of the need to observe and record is generally held to have set the pattern for later scientific voyages. Certainly Cook was convinced of the advantages of having artists on board. Even after Banks withdrew from the second voyage Cook arranged to employ other draughtsmen, 'for the express purpose' he later wrote of John Webber's appointment, 'of supplying the unavoidable imperfections of written accounts'. Webber was draughtsman to Cook's third and last expedition; his was the most complete pictorial illustration of any of the voyages and is well represented in the British Library's fine collection of Cook material (**68, 69**).

William Alexander, the first Keeper of Prints and Drawings at the British Museum, was another who made his reputation as an 'artist-traveller'. The drawings which he made in his official capacity as junior draughtsman to Lord Macartney's embassy to Peking in 1792–4 were to

70 'Pagoda at Lucknow, taken from Mr. Wombwell's house', 13 March 1786; sketchbook of Ozias Humphry, R.A. (1742–1810). *Pencil, black chalk and watercolour washes. 230 × 325 mm.* Add. MS 15962, f.19.

This is one of several drawings Humphry made at the house of John Wombwell, the East India Company's Accountant at Lucknow.

71 Self-portrait of Sir Robert
Ker Porter sketching a bas-
relief in Persia, 1817.

Watercolour. 425 × 545 mm. (detail).
Add. MS 14758, f.69.

stand him in good stead throughout his career and remain the work for
which he is best known. Macartney's was the first British embassy to
China. Though singularly unproductive in its commercial aims it did
bring back a wealth of new information about the Chinese country and
people which aroused considerable interest among the British public.
Alexander's sketches, worked up into delicate watercolours, were used for
engravings for both official and unofficial accounts. Two volumes of his
watercolours in the Manuscripts department were bequeathed by Sir John
Barrow of the Admiralty who had been Macartney's secretary. They contain the drawings used to illustrate Barrow's accounts of the voyage,
Travels in China (1804) and *A Voyage to Cochinchina* (1806) (**72, 74**).

For the Indian empire the natural place to look for the Library's best
topographical views is of course among the incomparable collections of the
India Office Library. Three stray sketchbooks of Indian material are,
however, preserved in the department of Manuscripts. The sketches are
those of Ozias Humphry, R. A., casual records of his three years in India
where he had gone in 1785 to paint miniatures at the courts of nabobs and
princes. Like other struggling artists of the 18th century Humphry saw
India as a possible avenue to wealth and fame, but when ill-health forced
him home in 1787 his success had not been great. Others enjoyed better
luck. Humphry visited Lucknow in 1786 (**70**) to find there Johan Zoffany

A Village and Cottagers

W. Alexander del.

Mandarin or Officer of State

W. Alexander del.

72 (*left*) Scenes in China; William Alexander (1767–1816). Album of drawings illustrating Lord Macartney's embassy to China, 1793–4.

Watercolour over pencil. 82 × 131 mm. (each drawing)
Add. MS 35300, f.23.

73 (*right*) 'Approach to the post of Derial in the Caucasus', 1817; Sir Robert Ker Porter (1777–1842).

Watercolour. 435 × 545 mm.
Add. MS 14758, f.12.

After the Russians annexed Georgia in 1783, Catharine the Great's favourite, Potemkin, built a fort at the northern end of the pass of Derial and an approach road along the ravine partly tunnelled through rock.

74 Aqueduct at Rio de Janeiro; William Alexander, 1805, from a drawing made 17 December 1792 on Lord Macartney's embassy to China.

Watercolour over pencil. 144 × 220 mm. Add. MS 35300, f.6.

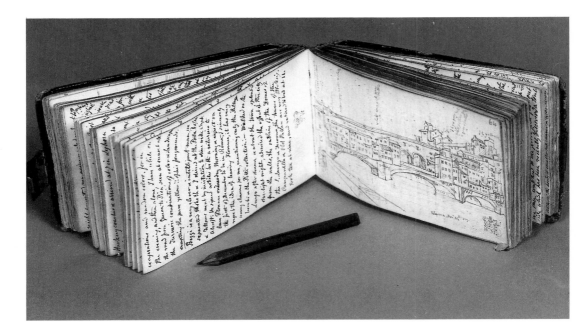

profitably at work on larger paintings of wealthy Company men and Indian courtiers.

Sir Robert Ker Porter, whose English sketchbook has already been mentioned (p. 43), is better remembered for his more adventurous travels. His appointment as an official painter to Tsar Alexander took him to Russia in 1805 and he subsequently married a Russian princess. He journeyed widely in Russia, Scandinavia, the Peninsula with Sir John Moore, Persia and Venezuela where he was the first British consul in Caracas. In 1821 he published a record of his travels in the Orient. The watercolour drawings from which the engravings to illustrate these volumes were taken are in the Library's collections (73). Ker Porter himself appears in a number of the pictures, dressed in the European military uniform which he was wont to adopt when travelling in Asiatic countries (71).

Looking very small and spontaneous beside Ker Porter's large pieces is a collection of pencil sketches made as part of a pocket travel journal by George Scharf the younger – many years before he became the first Director of the National Portrait Gallery. November 1839 found the young Scharf in Florence on his way through Italy into Asia Minor. Of course he drew the Ponte Vecchio (75); and he paused to reflect on the accuracy of the 'extravagant effects' which his illustrious fellow-artist Turner put into his landscapes: 'In the evening, and then alone, I have while on the road from Genoa to Pisa, seen at sunset all the Turneric combinations of color & landscape *excepting* the pure yellow!! of his foregrounds'.

By Victorian times parts of North Africa and the Middle East had become fashionable areas to tour for British artists and sightseers. But in

75 Ponte Vecchio, Florence, 24 November 1839; pocket travel diary kept by George Scharf the younger (1820–1895), on a journey to Asia Minor.

Pencil. 70 × 120 mm.
Add. MS 36488 A, ff.63v–64.

76 Ram Hormuz, Persia,
16 April 1879; from the
Arabian travel journals of
Lady Anne Blunt
(1837–1917).

Pencil and watercolour. 110 × 182 mm.
Add. MS 54049.

venturing in 1879 from Aleppo, down the valley of the Euphrates to the Persian Gulf, Wilfred Blunt and his wife, Lady Anne, granddaughter to Lord Byron, were risking a part of the Arab world which had been subjected to almost no European contact. From the expedition came at least two material results: horses bought in Arabia to form the famous Crabbet Arabian Stud at the Blunt home in Sussex; and Lady Anne's travel journal and sketches (**76**), later published to put a little more information about unknown Mesopotamia before the home reader. Lady Anne's watercolours are the simple, colourful, unassuming work of an amateur enthusiast whose aim is to capture the places she visited as authentically as she was able. In this respect she represented at the beginning of the age of the photograph much of the traditional spirit of topographical drawing.

Suggestions for further reading

An excellent general survey of topographical illustrations in public repositories is provided by M W Barley, *A Guide to British Topographical Collections* (1974). The standard and most comprehensive work on watercolours remains Martin Hardie, *Water-colour Painting in Britain* (3 vols, 1966–8); closely relevant in the present context are Michael Clarke, *The Tempting*

Prospect: A Social History of English Watercolours (1981) and Lindsay Stainton, *British Landscape Watercolours 1600–1860* (British Museum Publications, 1985) – one of a number of useful exhibition catalogues. P D A Harvey, *The History of Topographical Maps: Symbols, Pictures and Surveys* (1980) will be found helpful for the earlier period; and for the 16th and 17th centuries, E Croft-Murray and

High Street, Putney, Surrey;
George Sidney Shepherd
(1784–1862).

Watercolour heightened with white.
198 × 290 mm.
Printed Books, Crach.1. Tab. 1. b.1.
(vol. XX).

P Hulton, *Catalogue of British
Drawings in the British Museum*
(Vol. 1, 1960). R Russell, *Guide to
British Topographical Prints* (1979)
discusses the whole growth of the
print-making industry; Esther Moir,
The Discovery of Britain (1964) gives
a lively account of the fashion for
touring; and J Harris, *The Artist
and the Country House* (revised edn.
1985) is a lavishly illustrated history
of 'estate portraiture' from 1540 to
1870.

For individual artists a major
source of reference is H L
Mallalieu, *The Dictionary of British
Watercolour Artists up to 1920* (2nd
edn. 1986). Peter Jackson's recent
monograph, *George Scharf's
London: Sketches and Watercolours
of a Changing City, 1820–50* (1987)
also gives a vivid introduction to
London topography.

Readers who wish to study

drawings in the British Library
should first consult Barley (above),
which includes listings of BL
material. Drawings acquired before
the mid-19th century are covered by
the *Catalogue of the Manuscript
Maps, Charts & Plans and of the
Topographical Drawings in the
British Museum*, 2 vols. 1844
(Britain and France), one vol. 1861
(Foreign). After that date it is
necessary to use the general
catalogues of the department of
Manuscripts, for which see M A E
Nickson, *The British Library: Guide
to the catalogues and indexes of the
Department of Manuscripts* (2nd edn.
1982). The recently-published
User's Guide by Antony Griffiths
and Reginald Williams for the
*Department of Prints & Drawings in
the British Museum* (1987) has useful
sections on topography.